ROOTS AND RADICALS

BASICS

•**DEFINITION:** The real number **b** is the ***n*th** root of **a** if $b^n = a$.

•**RADICAL NOTATION:** If n≠0 then $a^{\frac{1}{n}} = \sqrt[n]{a}$ and $\sqrt[n]{a} = a^{\frac{1}{n}}$. The symbol $\sqrt{\ }$ is the radical or root symbol. The **a** is the radicand. The **n** is the index or order.

•**SPECIAL NOTE:** Equation $a^2 = 4$ has two solutions, **2 and -2**. However, the radical $\sqrt{a}$ represents only nonnegative square root of **a**.

•**DEFINITION OF SQUARE ROOT:** For any real number **a**, $\sqrt{a^2} = |a|$, that is, the nonnegative numerical value of **a** only; **EX:** $\sqrt{4} = +2$ only, by definition of the square root.

RULES

•**FOR ANY REAL NUMBERS, m AND n, WITH m/n IN LOWEST TERMS AND n≠0,**
$a^{\frac{m}{n}} = (a^m)^{\frac{1}{n}} = \sqrt[n]{a^m}$ **OR** $a^{\frac{m}{n}} = (a^{\frac{1}{n}})^m = (\sqrt[n]{a})^m$

•**FOR ANY REAL NUMBERS, m AND n, WITH m AND n, WITH m/n IN LOWEST TERMS AND n≠0,** $a^{-\frac{m}{n}} = 1/a^{\frac{m}{n}}$

•**FOR ANY NONZERO REAL NUMBER n,**
$(a^n)^{\frac{1}{n}} = a^1 = a$ **OR** $(a^{\frac{1}{n}})^n = a^1 = a$

•**FOR REAL NUMBERS a AND b AND NATURAL NUMBER n,**
$\sqrt[n]{ab} = (\sqrt[n]{a})(\sqrt[n]{b})$ **OR** $(\sqrt[n]{a})(\sqrt[n]{b}) = \sqrt[n]{ab}$

i.e., as long as the radical expressions have the **same index n**, they may be multiplied together and written as one radical expression of a product OR they may be separated and written as the product of two or more radical expressions; the radicands do not have to be the same for multiplication.

•**FOR REAL NUMBERS a AND b, AND NATURAL NUMBER n,**
$\frac{\sqrt[n]{a}}{\sqrt[n]{b}} = \sqrt[n]{\frac{a}{b}}$ **OR** $\sqrt[n]{\frac{a}{b}} = \frac{\sqrt[n]{a}}{\sqrt[n]{b}}$

i.e., as long as the radical expressions have the **same index** they may be written as one quotient under one radical symbol OR they may be separated and written as one radical expression over another radical expression; the radicands do not have to be the same for division.

•**TERMS CONTAINING RADICAL EXPRESSIONS CANNOT BE COMBINED**
Unless they are like or similar terms and the radical expressions which they contain are the same; the **indices and radicands must be the same for addition and subtraction**.

EXAMPLES: $3x\sqrt{2} + 5x\sqrt{2} = 8x\sqrt{2}$ but $3y\sqrt{5} + 7y\sqrt{3}$ cannot be combined because the radical expressions they contain are not the same. The terms $7m\sqrt{2}$ and $8m\sqrt[3]{2}$ cannot be combined because the indices (plural of index) are not the same.

OPERATIONS:

•**ADDITION AND SUBTRACTION:**
Only radical expressions which have the same index and the same radicand may be added.

•**MULTIPLICATION AND DIVISION:**
1. Monomials may be multiplied when the **indices are the same** even though the radicands are not.
EX: $(3x\sqrt{5})(2\sqrt{7}) = 6x\sqrt{35}$
2. Binomials may be multiplied using any method for multiplying regular binomial expressions if indices are the same. (e.g. **FOIL**). EX: $(9m + 2\sqrt{5})(3m - 5\sqrt{7}) = 27m^2 - 45m\sqrt{7} + 6m\sqrt{5} - 10\sqrt{35}$
3. Other polynomials are multiplied using the distributive property for multiplication if the indices are the same.
4. Division may be simplification of radical expressions or multiplication by the reciprocal of the divisor. Rationalize the answer so it is in lowest terms without a radical expression in the denominator.

U.S.$4.95
CAN. $7.50

January 2002

6 54614 20503 2

SIMPLIFYING RADICAL EXPRESSIONS

•**WHEN THE RADICAL EXPRESSION CONTAINS ONE TERM AND NO FRACTIONS (EX.** $\sqrt{12m^7}$**) THEN**
1. Take the greatest root of the coefficient; EXAMPLE: For $\sqrt{32}$ use $\sqrt{16}\cdot\sqrt{2}$, not $\sqrt{4}\cdot\sqrt{8}$ because $\sqrt{8}$ is not in simplest form.
2. Take the greatest root of each variable in the term. Remember $\sqrt[n]{a^n} = a$; that is, the power of the variable is divided by the index.
   a. This is accomplished by first noting if the power of the variable in the radicand is less than the index. If it is, the radical expression is in its simplest form.
   b. If the power of the variable is not less than the index, divide the power by the index. The quotient is the new power of the variable to be written outside of the radical symbol. The remainder is the new power of the variable still written inside of the radical symbol.
   EXAMPLES: $\sqrt[3]{a^7} = a^2\sqrt[3]{a}$; $\sqrt[3]{8ab^5} = 2b\sqrt[3]{ab^2}$

•**WHEN THE RADICAL EXPRESSION CONTAINS MORE THAN ONE TERM AND NO FRACTIONS** ($\sqrt{x^2+6x+9}$) **THEN**
1. **Factor**, if possible, and take the root of the **factors**. Never take the root of individual terms of a radical.
EX: $\sqrt{x^2+4} \neq x+2$, but $\sqrt{x^2+4x+4} = \sqrt{(x+2)^2} = x+2$ because the root of the factors $(x+2)^2$ was taken to get **x + 2** as the answer.
2. If the radicand is not factorable then the radical expression cannot be simplified because you cannot take the root of the terms of a radicand.

• **WHEN THE RADICAL EXPRESSION CONTAINS FRACTIONS**
1. If the fraction(s) is part of one radicand (under the radical symbol; Example: $\sqrt{\frac{2}{3}}$) then:
   a. Simplify the radicand as much as possible to make the radicand one rational expression so it can be separated into the root of the numerator over the root of the denominator. Simplify the radical expression in the numerator. Simplify the radical expression in the denominator.
   b. **Never leave a radical expression in the denominator.** It is not considered completely simplified until the fraction is in lowest terms. **Rationalize** the expression to remove the radical expression from the denominator. as follows:
      i. Step 1: **Multiply** the numerator and the denominator by the radical expression needed to eliminate the radical expression from the denominator. A radical expression in the numerator is acceptable. EX: $\frac{(5x\sqrt{2})}{(7\sqrt{3})}$ must be multiplied by $\frac{\sqrt{3}}{\sqrt{3}}$ so the denominator becomes 21 with no radical symbols in it. The numerator becomes $5x\sqrt{6}$.
      ii. Step 2: Write the answer in **lowest terms**.
2. If the fraction contains monomial radical expressions: EX: $\frac{(x)}{(\sqrt{5})}$, then
   a. If the radical expression is in the numerator only, simplify it and write the fraction in lowest terms.
   b. If the radical expression is in the denominator only, rationalize the fraction so no radical symbols remain there. Simplify the resulting fraction to lowest terms.
   c. If radical expressions are in both numerator and denominator, either:
      i. Simplify each separately, rationalize the denominator and write the answer in lowest terms, **OR**
      ii. Make the indices on all radical symbols the same, put the numerator and the denominator under one common radical symbol, write in lowest terms, separate again into a radical expression in the numerator and a radical expression in the denominator, rationalize the denominator and write the answer in lowest terms.
3. If the radical expressions are part of polynomials in a rational expression: [illegible] then:
   a. If [illegible] si[illegible]
   b. If [illegible] ra[illegible] d[illegible] C[illegible] M[illegible] 3[illegible] th[illegible]

9 781572 225039

RADICAL EXPRESSIONS IN EQUATIONS

•**RULE:** ***BOTH SIDES OF AN EQUATION MAY BE RAISED TO SAME POWER.*** Caution: Since both entire sides must be raised to the same power, place each side in a separate set of parentheses first.

•**STEPS**
1. If the equation has only one radical expression (EXAMPLE: $\sqrt{3x} + 5 = x$) then
   a. **Isolate** the radical expression on one side of the equal sign.
   b. **Raise** both sides of the equation to the same power as the index
   c. **Solve** the resulting equation.
   d. **Check** the solution(s) in the original equation because extraneous solutions are possible.
   EXAMPLE: $\sqrt{3x} + 5 = x$ becomes $\sqrt{3x} = x - 5$, then squaring both sides gives $3x = x^2 - 10x + 25$ because when the entire right side which is a binomial, **x - 5**, is squared, $(x - 5)^2$, the result is $x^2 - 10x + 25$. This is now a second degree equation. The steps for solving a quadratic equation should now be followed.
2. With equations containing two radical expressions: Ex: $\sqrt{x} + \sqrt{2x} = 4$
   a. **Change** the radical expressions to have the same index.
   b. **Separate** the radical expressions, placing one on each side of the equal sign.
   c. **Raise** both sides of the equation to the same power as the index.
   d. **Repeat** steps b and c above until all radical expressions are eliminated.
   e. **Solve** the resulting equation and check the solution(s).
3. If the equation has more than two radicals
   a. **Change** the radical expression to the same index.
   b. **Separate** as many radical expressions as possible on different sides of the equal sign.
   c. **Raise** both sides of the equation to the power of the index.
   d. **Repeat** steps b and c above until all radical expressions have been eliminated.
   e. **Solve** the resulting equation and check the solution(s).

QUADRATIC EQUATIONS

•**DEFINITION**
Second-degree equations in one variable which can be written in the form $ax^2+bx+c=0$ where **a**, **b**, and **c** are real numbers and **a≠0**

•**PROPERTY**
If **a** and **b** are real numbers and **(a)(b) =0** then either **a = 0** or **b = 0** or both equal zero. At least one of the numbers has to be equal to zero.

•**STEPS**
1. Set the equation **equal to zero**. Combine like terms. Write in descending order.
2. **Factor**; (if factoring is not possible then go to step 3)
   a. **Set each factor equal to zero**. See above: " if a product is equal to zero at least one of the factors must be zero."
   b. **Solve** each resulting equation and check the solution(s).
3. Use the **quadratic formula** if factoring is not possible.
   a. The quadratic formula is: $x = \frac{-b \pm \sqrt{b^2 - 4ac}}{2a}$
   b. **a**, **b**, and **c** come from the second-degree equation which is to be solved. After the second-degree equation has been set equal to zero, **a** is the coefficient (number in front of) of the second-degree term, **b** is the coefficient (number in front of) of the first-degree term (if no first-degree term is present then **b** is zero), and **c** is the constant term (no variable showing). **Note**: $ax^2 + bx + c = 0$
   c. Substitute the numerical values for **a**, **b**, and **c** into the quadratic formula.
   d. Simplify completely.
   e. Write the two answers, one with + in front of the radical expression in the formula, and one with - in front of the radical expression in the formula. Complete any additional simplification to get the answers in the required form.

COMPLEX NUMBERS

•**DEFINITION:** The set of all numbers, a + b*i*, where a and b are real numbers and $i^2 = -1$; that is, *i* is the number whose square equals -1 [illegible]
•**NOTE:** $i^2 = -1$ will be used in multi[illegible]nd division of complex numbers. The con[illegible]er $3 + 2i \neq 3 - 2i$ because the numbers mus[illegible]al to be equal.

•**OPERATIONS:**
1. **Addition and Subtraction:**
   a. Combine complex numbers as though the *i* [illegible]le; EX: $(4 + 5i) + (7 - 3i) = 11 + 2i$; $(-3 + 7i)$ - ([illegible] 5*i*
   b. The sum or difference of complex num[illegible]er complex number. Even the number 2[illegible]x number of the form a + b*i* where a = 21 [illegible]
2. **Multiplication and Division:**
   a. Multiply complex numbers using the met[illegible] plying two binomials. Remember that [illegible]e answer is not complete until i^2 has bee[illegible] -1 and simplified. EX: $(-3 + 5i)(1 - i)$ = -[illegible] $= -3 + 3i + 5i - 5(-1) = -3 + 3i + 5i + 5$ = [illegible]
   b. Divide complex numbers by rationalizi[illegible]i-nator. The answer is complete when the[illegible]al expression or *i* in the denominator and th[illegible] in simplest form. EX: The conjugate of the complex number $-3 + 12i$ is $-3-12i$.

Quick Study® ACADEMIC

ALGEBRA - PART 1

properties.dividing.equations.factoring

PARTS 1 & 2 COMBINED COVER THE BASIC PRINCIPLES OF ALGEBRA FOR INTERMEDIATE AND COLLEGE ALGEBRA COURSES

SET THEORY

NOTATION

{ } braces indicate the beginning and end of a set notation; when listed, elements or members must be separated by commas; **EX.** A = {4, 8, 16}; sets are finite (ending, or having a last element) unless otherwise indicated.

... indicates continuation of a pattern; **EX.** B = {5, 10, 15, ... , 85, 90}

... at the end indicates an infinite set, that is, a set with no last element; **EX.** C = {3, 6, 9, 12, ...}

| is a symbol which literally means "such that"

∈ means "is a member of" OR "is an element of"; **EX.** If A = {4, 8, 12} then 12∈A because 12 is in set A.

∉ means "is not a member of" OR "is not an element of"; **EX.** If B = {2, 4, 6, 8} then 3∉B because 3 is not in set B.

Ø empty set OR null set - a set containing no elements or members, but which is a subset of all sets; also written as { }.

⊂ means "is a subset of"; also may be written as ⊆

⊄ means "is not a subset of"; also may be written as ⊈

A ⊂B indicates that every element of set A is also an element of set B; **EX.** If A = {3, 6} and B = {1, 3, 5, 6, 7, 9} then A ⊂B because the 3 and 6 which are in set A are also in set B.

2^n is the number of subsets of a set when n equals the number of elements in that set; **EX.** If A = {4, 5, 6} then set A has 8 subsets because A has 3 elements and $2^3 = 8$.

OPERATIONS

∪ union

A∪B indicates the union of set A with set B; every element of this set is either an element of set A **OR** an element of set B; that is, to form the union of two sets put all of the elements of both sets together into one set making sure not to write any element more than once; **EX.** If A = {2,4} and B = {4, 8, 16} then A ∪ B = {2, 4, 8, 16}.

∩ intersection

A∩B indicates the intersection of set A with set B; every element of this set is also an element of **BOTH** set A and set B; that is, to form the intersection of two sets list only those elements which are found in **BOTH** of the two sets; **EX.** If A = {2,4} and B = {4, 8, 16} then A ∩ B = {4}.

$\overline{A}$ indicates the complement of set A; that is, all elements in the universal set which are **NOT** in set A; **EX.** If the Universal set is the set of Integers and A = {0, 1, 2, 3, ...} then $\overline{A}$ {-1, -2, -3, -4,...}. **A∩$\overline{A}$=Ø**

PROPERTIES

A = B all of the elements in set A are also in set B and all elements in set B are also in set A, although they do not have to be in the same order; **EX.** If A = {5, 10} and B = {10, 5} then A = B.

n(A) indicates the number of elements in set A; **EX.** If A = {2, 4, 6} then **n**(A) = 3.

~ means "is equivalent to"; that is, set A and set B have the same number of elements although the elements themselves may or may not be the same; **EX.** If A = {2, 4, 6} and B = {6, 12, 18} then A ~ B because **n**(A) = 3 and **n**(B) = 3.

A∩B =Ø indicates disjoint sets which have no elements in common.

SETS OF NUMBERS

Natural or Counting numbers = {1, 2, 3, 4, 5, ..., 11, 12, ...}

Whole numbers = {0, 1, 2, 3,..., 10, 11, 12, 13, ...}

Integers = {... , -4, -3, -2, -1, 0, 1, 2, 3, 4, ...}

Rational numbers = {p/q | p and q are integers, q≠0}; the sets of Natural numbers, Whole numbers, and Integers, as well as numbers which can be written as proper or improper fractions, are all subsets of the set of Rational numbers.

Irrational numbers = {x | x is a Real number but is not a Rational number}; the sets of Rational numbers and Irrational numbers have no elements in common and are therefore disjoint sets.

Real numbers = {x | x is the coordinate of a point on a number line}; the union of the set of Rational numbers with the set of Irrational numbers equals the set of Real numbers.

Imaginary numbers = {a***i*** | a is a real number and ***i*** is the number whose square is -1}; $i^2 = -1$; the sets of Real numbers and Imaginary numbers have no elements in common and are therefore disjoint sets.

Complex numbers = {a + b***i*** | a and b are real numbers and ***i*** is the number whose square is -1}; the set of Real numbers and the set of Imaginary numbers are both subsets of the set of Complex numbers; **EXs:** 4 + 7***i*** ; 3 - 2***i***

PROPERTIES OF REAL NUMBERS:
FOR ANY REAL NUMBERS a, b, AND c

PROPERTY	FOR ADDITION	MULTIPLICATION
Closure	a + b is a real number	ab is a real number
Commutative	a + b = b + a	ab = ba
Associative	(a + b) + c = a + (b + c)	(ab)c = a(bc)
Identity	0 + a = a and a + 0 = a	a·1 = a and 1·a = a
Inverse	a + (-a) = 0 and (-a) + a = 0	a · 1/a = 1 and 1/a · a = 1 if a ≠ 0
Distributive Property a(b + c) = ab + ac ; a(b - c) = ab - ac		

PROPERTIES OF EQUALITY:
FOR ANY REAL NUMBERS a, b, AND c

Reflexive	a = a
Symmetric	If a = b then b = a.
Transitive	If a = b and b = c then a = c.
Addition Property	If a = b then a + c = b + c.
Multiplication Property	If a = b then ac = bc.
Mult. Prop. of Zero	a · 0 = 0 and 0 · a = 0
Double Negative Property	-(-a) = a

PROPERTIES OF INEQUALITY
FOR ANY REAL NUMBERS a, b, AND c

Trichotomy Either a > b, or a = b, or a < b.

Transitive If a < b, and b < c then a < c.

Addition Property of Inequalities
If a < b then a + c < b + c.
If a > b then a + c > b + c.

Multiplication Property of Inequalities
If c ≠ 0 and c > 0, and a > b then ac > bc; also, if a < b then ac < bc.
If c ≠ 0 and c < 0, and a > b then ac < bc; also, if a < b then ac > bc.

OPERATIONS OF REAL NUMBERS

ABSOLUTE VALUE
|x| = x if x is zero or a positive number; |x| = -x if x is a negative number; that is, the distance (which is always positive) of a number from zero on the number line is the absolute value of that number;
EXs: |-4| = -(-4) = 4; |29| = 29; |0| = 0; |-43| = -(-43) =43

ADDITION
If the **signs** of the numbers are the **same: add** the absolute values of the numbers; the sign of the answer is the same as the signs of the original two numbers
EXs: -11 + -5 = -16 and 16 + 10 = 26
If the **signs** of the numbers are **different: subtract** the absolute values of the numbers; the answer has the same sign as the number with the larger absolute value;
EXs: -16 + 4 = -12 and -3 + 10 = 7.

SUBTRACTION
a - b = a + (-b); **subtraction is changed to addition of the opposite number**; that is, change the sign of the second number and follow the rules of addition (never change the sign of the first number since it is the number in back of the subtraction sign which is being subtracted; 14-6 ≠ -14 + -6;
EXs: 15 - 42 = 15 + (-42) = -27; -24 - 5 = -24 + (-5) = -29; -13 - (-45) = -13 + (+45) = 32; -62 - (-20) = -62 + (+20) = -42

MULTIPLICATION
The product of two numbers which have the **same** signs is **positive;**
EXs: (55)(3) = 165; (-30)(-4) = 120; (-5)(-12) = 60
The product of two numbers which have **different** signs is **negative** no matter which number is larger
EXs: (-3)(70) = -210; (21)(-40) = -840; (50)(-3) = -150

DIVISION (divisors do not equal zero)
The quotient of two numbers which have the **same sign** is **positive**;
EXs: (-14) / (-7) = 2; (44) / (11) = 4; (-4) / (-8) = .5
The quotient of two numbers which have **different signs** is **negative** no matter which number is larger;
EXs: (-24) / (6) = -4; (40) / (-8) = -5; (-14) / (56) = -.25

DOUBLE NEGATIVE
- (-a) = a; that is, the negative sign changes the sign of the contents of the parentheses; **EXs:** -(-4) = 4; -(-17) = 17.

ALGEBRAIC TERMS

COMBINING (ADDING OR SUBTRACTING)LIKE TERMS

a + a = 2a; when adding or subtracting terms they must have exactly the same variables and exponents although not necessarily in the same order; these are called like terms. The coefficients (numbers in the front) may or may not be the same.

RULE: combine (add or subtract) only the coefficients of like terms and never change the exponents during addition or subtraction;
EXs: $4xy^3$ **and** $-7y^3x$ are like terms and may be combined in this manner: $4xy^3 + -7y^3x = -3xy^3$. Notice only the coefficients were combined and no exponent changed. $-15a^2bc$ and $3bca^4$ are not like terms because the exponents of the **a** are not the same in both terms, so they may not be combined.

MULTIPLYING

PRODUCT RULE FOR EXPONENTS
$(a^m)(a^n) = a^{m+n}$; any terms may be multiplied, not just like terms. The coefficients and the variables are multiplied which means the exponents also change;
RULE: multiply the coefficients and multiply the variables (this means you have to add the exponents of the same variable);
EX: $(4a^2c)(-12a^3b^2c) = -48a^5b^2c^2$, notice that 4 times -12 became -48, a^2 times a^3 became a^5, c times c became c^2, and the b^2 was written down.

DISTRIBUTIVE PROPERTY FOR POLYNOMIALS
Type 1: $a(c + d) = ac + ad$; **EX:** $4x^3(2xy + y^2) = 8x^4y + 4x^3y^2$
Type 2: $(a + b)(c + d) = a(c + d) + b(c + d) = ac + ad + bc + bd$;
EX: $(2x + y)(3x - 5y) = 2x(3x - 5y) + y(3x - 5y) = 6x^2 - 10xy + 3xy - 5y^2 = 6x^2 - 7xy - 5y^2$
Type 3: $(a + b)(c^2 + cd + d^2) = a(c^2 + cd + d^2) + b(c^2 + cd + d^2) = ac^2 + acd + ad^2 + bc^2 + bcd + bd^2$
EX: $(5x+3y)(x^2-6xy+4y^2) = 5x(x^2-6xy+4y^2) + 3y(x^2 - 6xy + 4y^2) = 5x^3 - 30x^2y + 20xy^2 + 3x^2y - 18xy^2 + 12y^3 = 5x^3 - 27x^2y + 2xy^2 + 12y^3$

FOIL METHOD FOR PRODUCTS OF BINOMIALS
This is a popular method for multiplying 2 terms by 2 terms only. FOIL means first times first, outer times outer, inner times inner, and last times last.
EX: $(2x+3y)(x+5y)$ would be multiplied by multiplying first term times first term, 2x times x = $2x^2$; outer term times outer term, 2x times 5y = 10xy; inner term times inner term, 3y times x = 3xy; and last term times last term, 3y times 5y = $15y^2$; then combining the like terms of 10xy and 3xy gives 13xy with the final answer equaling $2x^2 + 13xy + 15y^2$

SPECIAL PRODUCTS
Type 1: $(a + b)^2 = (a + b)(a + b) = a^2 + 2ab + b^2$
Type 2: $(a - b)^2 = (a - b)(a - b) = a^2 - 2ab + b^2$
Type 3: $(a + b)(a - b) = a^2 + 0ab - b^2 = a^2 - b^2$

EXPONENT RULES
Rule 1: $(a^m)^n = a^{m \cdot n}$; $(a^m)^n$ means the parentheses contents are multiplied n times and when you multiply you add exponents;
EX: $(-2m^4n^2)^3 = (-2m^4n^2)(-2m^4n^2)(-2m^4n^2) = -8m^{12}n^6$, notice the parentheses were multiplied 3 times and then the rules of regular multiplication of terms were used;
SHORTCUT RULE: when raising a term to a power just multiply exponents;
EX: $(-2m^4n^2)^3 = -2^3m^{12}n^6$, notice the exponents of the -2, m^4 and n^2 were all multiplied by the exponent 3, and that the answer was the same as the example above.
CAUTION: $-a^m \neq (-a)^m$; these two expressions are different;
EXs: $-4yz^2 \neq (-4yz)^2$ because $(-4yz)^2 = (-4yz)(-4yz) = 16y^2z^2$ while $-4yz^2$ means $-4 \cdot y \cdot z^2$ and the exponent 2 applies only to the z in this situation.
Rule 2: $(ab)^m = a^m b^m$; **EX:** $(6x^3y)^2 = 6^2x^6y^2 = 36x^6y^2$
BUT $(6x^3 + y)^2 = (6x^3 + y)(6x^3 + y) = 36x^6 + 12x^3y + y^2$; because there is more than one term in the parentheses the distributive property for polynomials must be used in this situation.
Rule 3: $\left(\frac{a}{b}\right)^m = \frac{a^m}{b^m}$ **when b ≠ 0 ; EX:** $\left(\frac{-4x^2y}{5z}\right)^2 = \frac{16x^4y^2}{25z^2}$
Rule 4: Zero Power $a^0 = 1$ when a ≠ 0

DIVIDING

QUOTIENT RULE: ***$a^m / a^n = a^{m-n}$; any terms may be divided, not just like terms; the coefficients and the variables are divided which means the exponents also change;***
RULE: divide the coefficients and divide the variables (this means you have to subtract the exponents of matching variables);
EX: $(-20x^5y^2z) / (5x^2z) = -4x^3y^2$, notice that -20 divided by 5 became -4, x^5 divided by x^2 became x^3, and z divided by z became one and therefore did not have to be written because 1 times $-4x^3y^2$ equals $-4x^3y^2$.

NEGATIVE EXPONENT: $a^{-n} = 1 / a^n$ when a ≠ 0;
EXs: $2^{-1} = 1/2$; $(4z^{-3}y^2) / (-3ab^{-1}) = (4y^2b^1) / (-3az^3)$ (notice that the 4 and the -3 both stayed where they were because they both had an invisible exponent of positive 1; the **y** remained in the numerator and the **a** remained in the denominator because their exponents were both positive numbers; the **z** moved down and the **b** moved up because their exponents were both negative numbers).

ORDER OF OPERATIONS

FIRST, *SIMPLIFY ANY ENCLOSURE SYMBOLS: PARENTHESES ( ), BRACKETS [ ], BRACES { } IF PRESENT:*

a. Work the enclosure symbols from the innermost and work outward.

b. Work separately above and below any fraction bars since the entire top of a fraction bar is treated as though it has its own invisible enclosure symbols around it and the entire bottom is treated the same way.

SECOND, *SIMPLIFY ANY EXPONENTS AND ROOTS, WORKING FROM LEFT TO RIGHT;*

Note: The $\sqrt{\ }$ symbol is used only to indicate the positive root, except that $\sqrt{0} = 0$.

THIRD, *DO ANY MULTIPLICATION AND DIVISION IN THE ORDER IN WHICH THEY OCCUR, WORKING FROM LEFT TO RIGHT;*

Note: If division comes before multiplication then it is done first, if multiplication comes first then it is done first.

FOURTH, *DO ANY ADDITION AND SUBTRACTION IN THE ORDER IN WHICH THEY OCCUR, WORKING FROM LEFT TO RIGHT;*

Note: If subtraction comes before addition in the problem then it is done first, if addition comes first then it is done first.

STEPS FOR SOLVING A FIRST DEGREE EQUATION WITH ONE VARIABLE

FIRST, *ELIMINATE ANY FRACTIONS BY USING THE MULTIPLICATION PROPERTY OF EQUALITY*

EX: $1/2\,(3a + 5) = 2/3\,(7a - 5) + 9$ would be multiplied on both sides of the = sign by the lowest common denominator of 1/2 and 2/3, which is 6; the result would be $3(3a + 5) = 4(7a - 5) + 54$, notice that only the 1/2, the 2/3, and the 9 were multiplied by 6 and not the contents of the parentheses; the parentheses will be handled in the next step which is distribution;

SECOND, *SIMPLIFY THE LEFT SIDE OF THE EQUATION AS MUCH AS POSSIBLE BY USING THE ORDER OF OPERATIONS, THE DISTRIBUTIVE PROPERTY, AND COMBINING LIKE TERMS. DO THE SAME TO THE RIGHT SIDE OF THE EQUATION;*

EX: Use distribution first, $3(2k - 5) + 6k - 2 = 5 - 2(k + 3)$ would become $6k - 15 + 6k - 2 = 5 - 2k - 6$ and then combine like terms to get $12k - 17 = -1 - 2k$.

THIRD, *APPLY THE ADDITION PROPERTY OF EQUALITY TO SIMPLIFY AND ORGANIZE ALL TERMS CONTAINING THE VARIABLE ON ONE SIDE OF THE EQUATION AND ALL TERMS WHICH DO NOT CONTAIN THE VARIABLE ON THE OTHER SIDE;*

EX: $12k - 17 = -1 - 2k$ would become $2k + 12k - 17 + 17 = -1 + 17 - 2k + 2k$ and then combining like terms, $14k = 16$.

FOURTH, *APPLY THE MULTIPLICATION PROPERTY OF EQUALITY TO MAKE THE COEFFICIENT OF THE VARIABLE 1;*

EX: $14k = 16$ would be multiplied on both sides by 1/14 (or divided by 14) to get a 1 in front of the k so the equation would become $1k = 16/14$ or simply $k = 1\ 1/7$ or 1.143;

FIFTH, *CHECK THE ANSWER BY SUBSTITUTING IT FOR THE VARIABLE IN THE ORIGINAL EQUATION TO SEE IF IT WORKS.*

NOTE:
1. Some equations have exactly one solution (answer). They are **conditional equations** (EX: $2k = 18$).
2. Some equations work for all real numbers. They are **identities** (EX: $2k = 2k$).
3. Some equations have no solutions. They are **inconsistent equations** (EX: $2k + 3 = 2k + 7$).

SOLVING A FIRST DEGREE INEQUALITY WITH ONE VARIABLE

ADDITION PROPERTY OF INEQUALITIES

For all real numbers a, b, and c, the inequalities $a < b$ and $a + c < b + c$ are equivalent; that is, any terms may be added to both sides of an inequality and the inequality remains a true statement. This **also applies to $a > b$ and $a + c > b + c$**.

MULTIPLICATION PROPERTY OF INEQUALITIES

1. **For all real numbers a, b, and c, with $c \neq 0$ and $c > 0$, the inequalities $a > b$ and $ac > bc$ are equivalent and the inequalities $a < b$ and $ac < bc$ are equivalent**; that is, when c is a positive number the inequality symbols stay the same as they were before the multiplication;
   EX: If $8 > 3$ then multiplying by 2 would make $16 > 6$, which is a true statement.
2. **For all real numbers a, b, and c, with $c \neq 0$ and $c < 0$, the inequalities $a > b$ and $ac < bc$ are equivalent and the inequalities $a < b$ and $ac > bc$ are equivalent**; that is, when c is a negative number the inequality symbols must be reversed from the way they were before the multiplication for the inequality to remain a true statement;
   EX: If $8 > 3$ then multiplying by -2 would make $-16 > -6$, which is false unless the inequality symbol is reversed to make it true, $-16 < -6$.

FACTORING

Some algebraic polynomials cannot be factored. The following are methods of handling those which can be factored. When the factoring process is complete the answer can always be checked by multiplying the factors out to see if the original problem is the result. That will happen if the factorization is a correct one.

A polynomial is factored when it is written as a product of polynomials with integer coefficients and all of the factors are prime. The order of the factors does not matter.

FIRST STEP - 'GCF'

FACTOR OUT THE GREATEST COMMON FACTOR (GCF), IF THERE IS ONE

The **GCF** is the largest number which will divide evenly into every coefficient together with the lowest exponent of each variable common to all terms. **EX:** $15a^3c^3 + 25a^2c^4d - 10a^2c^3d$ has a greatest common factor of $5a^2c^3$ because 5 divides evenly into 15, 25, and 10; the lowest degree of **a** in all three terms is 2; the lowest degree of **c** is 3; the GCF is $5a^2c^3$; the factorization is $5a^2c^3\,(3a + 5cd - 2d)$

SECOND STEP - CATEGORIZE AND FACTOR

The 2nd step of factoring is to identify the problem as belonging in one of the following categories. Be sure to place the terms in the correct order first: highest degree term to the lowest degree term. EX: $-2A^3 + A^4 + 1 = A^4 - 2A^3 + 1$

CATEGORY	FORM OF PROBLEM	FORM OF FACTORS
TRINOMIALS *(3 TERMS)*	$ax^2 + bx + c$ *($a \neq 0$)*	**If $a = 1$: $(x + h)(x + k)$ where $h \bullet k = c$ and $h + k = b$; h and k may be either positive or negative numbers. If $a \neq 1$: $(mx + h)(nx + k)$ where $m \bullet n = a$, $h \bullet k = c$, and $h \bullet n + m \bullet k = b$; m, h, n and k may be either positive or negative numbers. Trial and error methods may be needed.** *(see Special Hints below)*
	$x^2 + 2cx + c^2$ *(perfect square)*	**$(x + c)(x + c) = (x + c)^2$ where c may be either a positive or a negative number**
BINOMIALS *(2 TERMS)*	$a^2x^2 - b^2y^2$ *(difference of 2 squares)*	$(ax + by)(ax - by)$
	$a^2x^2 + b^2y^2$ *(sum of 2 squares)*	PRIME -- can not be factored!
	$a^3x^3 + b^3y^3$ *(sum of 2 cubes)*	$(ax + by)(a^2x^2 - abxy + b^2y^2)$
	$a^3x^3 - b^3y^3$ *(difference of 2 cubes)*	$(ax - by)(a^2x^2 + abxy + b^2y^2)$ *(see Special Hints below)*
PERFECT CUBES (4 TERMS)	$a^3x^3 + 3a^2bx^2 + 3ab^2x + b^3$ $a^3x^3 - 3a^2bx^2 + 3ab^2x - b^3$	$(ax + b)^3 = (ax + b)(ax + b)(ax + b)$ $(ax - b)^3 = (ax - b)(ax - b)(ax - b)$
GROUPING	$ax + ay + bx + by$ *(2-2 grouping)*	$a(x + y) + b(x + y) = (x + y)(a + b)$
	$x^2 + 2cx + c^2 - y^2$ *(3-1 grouping)*	$(x + c)^2 - y^2 = (x + c + y)(x + c - y)$
	$y^2 - x^2 - 2cx - c^2$ *(1-3 grouping)*	$y^2 - (x + c)^2 = (y + x + c)(y - x - c)$

SPECIAL HINTS

TRINOMIALS: WHERE THE COEFFICIENT OF THE HIGHEST DEGREE TERM IS NOT 1

The first term in each set of parentheses must multiply to equal the first term (highest degree) of the problem. The second term in each set of parentheses must multiply to equal the last term in the problem. The middle term must be checked on a trial and error basis using: outer times outer plus inner times inner; $ax^2 + bx + c = (mx + h)(nx + k)$ where mx times nx equals ax^2, h times k equals c, and mx times k plus h times nx equals bx.

EX: To factor $3x^2 + 17x - 6$ all of the following are possible correct factorizations, $(3x + 3)(x - 2)$; $(3x + 2)(x - 3)$; $(3x + 6)(x - 1)$; $(3x + 1)(x - 6)$. However, the only set which results in a 17x for the middle term when applying outer times outer plus inner times inner is the last one, $(3x + 1)(x - 6)$. It results in $-17x$ and $+17x$ is needed, so both signs must be changed to get the correct middle term. Therefore, the correct factorization is $(3x - 1)(x + 6)$.

BINOMIALS: THE SUM OR DIFFERENCE OF TWO CUBES

This type of problem, $a^3x^3 \pm b^3y^3$, requires the memorization of the following procedure: The factors are two sets of parentheses with 2 terms in the first set and 3 terms in the second. To find the 2 terms in the first set of parentheses take the cube root of both of the terms in the problem and join them by the same middle sign found in the problem. The 3 terms in the second set of parentheses are generated from the 2 terms in the first set of parentheses. The first term in the second set of parentheses is the square of the first term in the first set of parentheses; the last term in the second set is the square of the last term in the first set; the middle term of the second set of parentheses is found by multiplying the first term and the second term from the first set of parentheses together and changing the sign.

Thus, $a^3x^3 \pm b^3y^3 = (ax \pm by)(a^2x^2 \mp abxy + b^2y^2)$

EX: To factor $27x^3 - 8$ find the first set of parentheses to be $(3x - 2)$ because the cube root of $27x^3$ is 3x and the cube root of 8 is 2. Find the 3 terms in the second set of parentheses by squaring 3x to get $9x^2$; square the last term -2 to get +4; and to find the middle term multiply 3x times -2 and change the sign to get +6x. Therefore, the final factorization of $27x^3 - 8$ is $(3x - 2)(9x^2 + 6x + 4)$.

PERFECT CUBES (4 TERMS)

Perfect cubes such as $a^3x^3 \pm 3a^2bx^2 + 3ab^2x \pm b^3$, factor into three sets of parentheses each containing exactly the same two terms; therefore, the final factorization is written as one set of parentheses to the third power, thus a perfect cube, $(ax \pm b)^3 = a^3x^3 \pm 3a^2bx^2 + 3ab^2x \pm b^3$

EX: To factor $27x^3 - 54x^2 + 36x - 8$ it must be first observed that the problem is in correct order and that it is a perfect cube; then the answer is simply the cube roots of the first term and the last term placed in a set of parentheses to the third power; so the answer to this example is $(3x - 2)^3$.

STEPS FOR SOLVING A FIRST DEGREE INEQUALITY WITH ONE VARIABE

FIRST, Simplify the left side of the inequality in the same manner as an equation, applying the order of operations, the distributive property, and combining like terms. **Simplify the right side in the same manner.**

SECOND, Apply the Addition Property of Inequality to get all terms which have the variable on one side of the inequality symbol and all terms which do not have the variable on the other side of the symbol.

THIRD, Apply the Multiplication Property of Inequality to get the coefficient of the variable to be 1; (remember to reverse the inequality symbol when multiplying or dividing by a negative number, this is NOT done when multiplying or dividing by a positive number).

FOURTH, Check the solution by substituting some numerical values of the variable in the original inequality.

RATIONAL EXPRESSIONS

Quotient of two polynomials where denominator cannot equal zero is a rational expression.

EX: $\frac{(x+4)}{(x-3)}$ where $x \neq 3$ since 3 would make denominator, x - 3, equal to zero.

BASICS

DOMAIN: *SET OF ALL REAL NUMBERS WHICH CAN BE USED TO REPLACE VARIABLE*

EX: The domain for the rational expression $\frac{(x+5)(x-2)}{(x+1)(4-x)}$ is $\{x \mid x \in \text{Reals and } x \neq -1 \text{ or } x \neq 4\}$

a. That is, **x** can be any real number except **-1** or **4** because **-1** makes $(x + 1)$ equal to zero and **4** makes $(4 - x)$ equal to zero; therefore, the denominator would equal zero, which it must not.

b. Notice that numbers which make numerator equal to zero, **-5** and **2**, are members of the domain since fractions may have zero in numerator but not in denominator.

•RULE 1

1. If **x/y** is a rational expression then **x/y = xa/ya** when **a≠0**

a. That is, you may multiply a rational expression (or fraction) by any nonzero value as long as you multiply both numerator and denominator by the same value.

i. Equivalent to multiplying by **1** since **a/a=1; (x/y)(1)=(x/y)(a/a) = xa/ya**

ii. Note: **1** is equal to any fraction which has the same numerator and denominator.

•RULE 2

1. If $\frac{xa}{ya}$ is a rational expression, $\frac{xa}{ya} = \frac{x}{y}$ when **a≠0**

a. That is, you may write a rational expression in lowest terms because $\frac{xa}{ya} = \left(\frac{x}{y}\right)\left(\frac{a}{a}\right) = \left(\frac{x}{y}\right)(1) = \frac{x}{y}$ since $\frac{a}{a} = 1$

•LOWEST TERMS

1. Rational expressions are in lowest terms when they have no common factors other than **1**.
2. Step 1: Completely **factor** both numerator & denominator.
3. Step 2: **Divide** both the numerator and the denominator by the greatest common factor or by the common factors until no common factors remain;

EX: $\frac{(x^2 + 8x + 15)}{(x^2 + 3x - 10)} = \frac{(x + 5)(x + 3)}{(x + 5)(x - 2)} = \frac{(x + 3)}{(x - 2)}$ because the common factor of $(x + 5)$ was divided into the numerator and the denominator since $\frac{(x+5)}{(x+5)} = 1$

4. NOTE: **Only factors** can be divided into both numerators and denominators — **never terms**.

SUBTRACTION

(Denominators must be the same)

•RULE 1

1. If $\frac{a}{b}$ and $\frac{c}{b}$ are rational expressions and **b≠o** then $\left(\frac{a}{b}\right) - \left(\frac{c}{b}\right) = \left(\frac{a}{b}\right) + \left(\frac{-c}{b}\right) = \frac{(a - c)}{b}$

a. If denominators are the same be sure to change all signs of the terms in numerator of rational expression which is behind (to right of) subtraction sign; then add numerators and write result over common denominator.

EXAMPLE: $\frac{x - 3}{x + 1} - \frac{x + 7}{x + 1} = \frac{x - 3 + (-x) + (-7)}{x + 1} = \frac{-10}{x + 1}$

•RULE 2

1. If $\frac{a}{b}$ and $\frac{c}{d}$ are rational expressions and $b \neq 0$ and $d \neq 0$, then $\frac{a}{b} - \frac{c}{d} = \frac{(ad)}{(bd)} - \frac{(cb)}{(bd)} = \frac{(ad - cb)}{(bd)}$

a. If denominators are not the same they must be made the same before numerators can be subtracted. Be sure to change signs of all terms in numerator of rational expression which follows subtraction sign after rational expressions have been made to have a common denominator. Combine numerator terms and write result over common denominator.

b. Note: When denominators of rational expressions are additive inverses (opposite signs) then signs of all terms in denominator of expression behind subtraction sign should be changed. This will make denominators the same and terms of numerators can be combined as they are. Subtraction of rational expressions is changed to addition of opposite of either numerator (most of time) or denominator (most useful when denominators are additive inverses) but never both.

•SUBTRACTION STEPS:

1. If the denominators are the same
   a. Change signs of all terms in numerator of a rational expression which follows any subtraction sign.
   b. Add the numerators.
   c. Write answer to this addition over common denominator.
   d. Write final answer in lowest terms, making sure to follow directions for finding lowest terms as indicated above.

   EX: $\frac{(x+2)}{(x-6)} - \frac{(x-1)}{(x-6)} = \frac{[x+2+(-x)+1]}{(x-6)} = \frac{3}{(x-6)}$

2. If the denominators are not the same
   a. Find the least common denominator.
   b. Change all of the rational expressions so they have the same common denominator.
   c. Multiply factors in the numerators if there are any.
   d. Change the signs of all of the terms in the numerator of any rational expressions which are behind subtraction signs.
   e. Add numerators.
   f. Write the sum over the common denominator.
   g. Write the final answer in lowest terms.

EX: $\frac{x+3}{x+5} - \frac{x+1}{x-1} = \frac{(x+3)(x-1)}{(x+5)(x-1)} - \frac{(x+1)(x+5)}{(x-1)(x+5)} = \frac{-4x - 8}{x^2 + 4x - 5}$

h. **NOTE**: If denominators are of degree greater than one, try to factor all denominators first so least common denominator will be product of all **different factors** from each denominator.

OPERATIONS

MULTIPLICATION

(Denominators do not have to be the same)

•RULE

1. If **a**, **b**, **c**, & **d** are real numbers and **b** and **d** are nonzero numbers, $\left(\frac{a}{b}\right)\left(\frac{c}{d}\right) = \frac{(ac)}{(bd)}$; [top times top and bottom times bottom]

•MULTIPLICATION STEPS

1. Completely **factor** all numerators and denominators.
2. **Write** problem as one big fraction with all numerators written as factors (multiplication indicated) on top and all denominators written as factors (multiplication indicated) on bottom.
3. **Divide** both numerator and denominator by all of the common factors; that is, write in lowest terms.
4. **Multiply** the remaining factors in the numerators together and write the result as the final numerator.
5. **Multiply** the remaining factors in the denominators together and write the result as the final denominator.

EX. $\left(\frac{x + 3}{x^2 + 2x + 1}\right)\left(\frac{x^2 - 2x - 3}{x^2 - 9}\right) = \frac{(x + 3)}{(x + 1)(x + 1)} \cdot \frac{(x - 3)(x + 1)}{(x + 3)(x - 3)} = \frac{1}{x + 1}$

DIVISION

•DEFINITION

1. Reciprocal of a rational expression $\frac{x}{y}$, is $\frac{y}{x}$ because $\left(\frac{x}{y}\right)\left(\frac{y}{x}\right) = 1$ [reciprocal may be found by inverting expression]

EXAMPLE: The reciprocal of $\frac{(x - 3)}{(x + 7)}$ is $\frac{(x + 7)}{(x - 3)}$

•RULE

1. If **a**, **b**, **c**, and **d** are real numbers and **a**, **b**, **c**, and **d** are non zero numbers, then $\frac{a}{b} \div \frac{c}{d} = \left(\frac{a}{b}\right)\left(\frac{d}{c}\right) = \frac{(ad)}{(bc)}$

•DIVISION STEPS

1. **Reciprocate** (flip) rational expression found behind division sign (immediately to right of division sign).
2. **Multiply** resulting rational expressions, making sure to follow steps for multiplication as listed above.

EX: $\frac{x^2 - 2x - 15}{x^2 - 10x + 25} \div \frac{(x + 2)}{(x - 5)} = \frac{x^2 - 2x - 15}{x^2 - 10x + 25} \times \frac{(x - 5)}{(x + 2)}$

Numerators and denominators would then be factored, written in lowest terms, and yield a final answer of $\frac{(x + 3)}{(x + 2)}$

ADDITION

(Denominators must be the same)

•RULE 1

1. If **a/b** and **c/b** are rational expressions and **b≠0**, then $\frac{a}{b} + \frac{c}{b} = \frac{(a + c)}{b}$

a. If denominators are already the same simply add numerators and write this sum over common denominator.

•RULE 2

1. If $\frac{a}{b}$ and $\frac{c}{d}$ are rational expressions and **b≠0** and **d≠0** then $\frac{a}{b} + \frac{c}{d} = \frac{(ad)}{(bd)} + \frac{(cb)}{(bd)} = \frac{(ad + cb)}{(bd)}$

a. If denominators are not the same they must be made the same before numerators can be added.

•ADDITION STEPS

1. If the denominators are the same
   a. Add the numerators.
   b. Write answer over common denominator.
   c. Write final answer in lowest terms, making sure to follow directions for finding lowest terms as indicated above.

   EXAMPLE: $\frac{(x + 2)}{(x - 6)} + \frac{(x - 1)}{(x - 6)} = \frac{(2x + 1)}{(x - 6)}$

2. If the denominators are not the same
   a. Find the least common denominator.
   b. Change all of the rational expressions so they have the same common denominator.
   c. Add numerators.
   d. Write the sum over the common denominator.
   e. Write the final answer in lowest terms. **EXAMPLE**:

   $\frac{x+3}{x+5} + \frac{x+1}{x-1} = \frac{(x+3)(x-1)}{(x+5)(x-1)} + \frac{(x+1)(x+5)}{(x-1)(x+5)} = \frac{2x^2 + 8x + 2}{x^2 + 4x - 5}$

   f. **NOTE**: If denominators are of degree greater than one, try to factor all denominators first so least common denominator will be the product of all **different factors** from each denominator.

COMPLEX FRACTIONS:

AN UNDERSTANDING OF 'OPERATIONS' SECTION IS REQUIRED TO WORK 'COMPLEX FRACTIONS'

•DEFINITION: *A rational expression having a fraction in the numerator or denominator or both is a complex fraction* EX: $\frac{x - \frac{1}{x}}{x}$

•TWO AVAILABLE METHODS:

1. **Simplify** numerator (combine rational expressions found only on top of the complex fraction) and denominator (combine rational expressions found only on bottom of the complex fraction) then divide numerator by denominator; that is, multiply numerator by reciprocal (flip) of denominator. **OR**
2. **Multiply** the complex fraction (both in numerator & denominator) by least common denominator of all individual fractions which appear anywhere in the complex fraction. This will eliminate the fractions on top & bottom of the complex fraction and result in one simple rational expression. Follow steps previously listed for simplifying rational expressions.

SYNTHETIC DIVISION

•DEFINITION

1. A process used to divide a polynomial by a binomial in the form of **x + h** where **h** is an integer.

•STEPS

1. Write polynomial in descending order [from highest to lowest power of variable]. **EX:** $3x^3 - 6x + 2$
2. Write all coefficients of dividend under long division symbol, making sure to write zeros which are coefficients of powers of variable which are not in polynomial.
   EX: Writing coefficients of polynomial in example above, write **3 0 -6 2** because a zero is needed for the x^2 since this power of **x** does not appear in polynomial and therefore has a coefficient of zero.
3. Write the binomial in descending order; **EX**: **x - 2**.
4. Write additive inverse of constant term of binomial in front of long division sign as divisor.
   EX: The additive inverse of the **-2** in the binomial **x -2** is simply **+2**; that is, change the sign of this term.
5. Bring up first number in dividend so it will become first number in quotient (the answer).
6. Multiply number just placed in quotient by divisor, **2**
   a. Add result of multiplication to next number in dividend
   b. Result of this addition is next number coefficient in quotient; so write it over next coefficient in dividend.
7. Repeat step 6 until all coefficients in dividend have been used
   a. Last number in the quotient is the numerator of a remainder which is written as a fraction with the binomial as the denominator **EX:** 2 ⟌3 0 - 6 2 results in a quotient of **3 6 6** with remainder **14**; therefore
   $(3x^3 - 6x + 2) \div (x - 2) = 3x^2 + 6x + 6 + \frac{14}{(x - 2)}$
8. First exponent in answer (quotient) is one less than highest power of dividend because division was by a variable to first degree.

RATIONAL EXPRESSIONS IN EQUATIONS

•DEFINITION:

*Equations which contain rational expressions, that is, **algebraic fractions**.*

•STEPS:

1. **Determine least common denominator** for all rational expressions in equation.
2. Use the multiplication property of equality to **multiply** all terms on both sides of the equality sign by the common demoninator and thereby **eliminate all algebraic fractions**.
3. **Solve** resulting equation using appropriate steps, depending on degree of equation which resulted from following step 2
4. **Check** answers because numerical values which cause denominators of rational expressions in original equation to be equal to zero are extraneous solutions, not true solutions of original equation.